SURPRISING SCIENCE PUZZLES

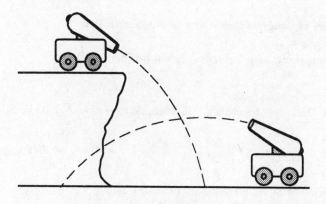

Erwin Brecher

Sterling Publishing Co., Inc.
New York

To Ellen and Cathy

Edited by Claire Bazinet

Library of Congress Cataloging-in-Publication Data

Brecher, Erwin.
    Surprising science puzzles / Erwin Brecher.
      p. cm.
    Includes index.
    ISBN 0-8069-0698-7
    1. Science—Miscellanea.   2. Scientific recreations. I. Title.
  Q173.B89 1994
  793.8—dc20   94-16838
                                            CIP

10 9 8 7 6 5 4 3 2 1

Published by Sterling Publishing Company, Inc.
387 Park Avenue South, New York, N.Y. 10016
© 1995 by Erwin Brecher
Distributed in Canada by Sterling Publishing
% Canadian Manda Group, One Atlantic Avenue, Suite 105
Toronto, Ontario, Canada M6K 3E7
Distributed in Great Britain and Europe by Cassell PLC
Wellington House, 125 Strand, London WC2R 0BB, England
Distributed in Australia by Capricorn Link (Australia) Pty Ltd.
P.O. Box 6651, Baulkham Hills, Business Centre, NSW 2153, Australia

Manufactured in the United States of America

Sterling ISBN 0-8069-0698-7

CONTENTS

Acknowledgments

The laws of physics are the creation of nature and not of man. However, this still leaves ample scope to construct around these laws intriguing science puzzles to challenge the cognitive abilities of any reader.

In compiling this book I have, in addition to my own material, enjoyed generous permission from other authors and their publishers. For this, I would like to express my appreciation to Scot Morris, Victor Serebriakoff (Octopus Publishing), Gyles Brandreth, Stephen Barr, and L. H. Longley-Cook. Their books are valuable additions to any puzzle library.

My thanks go, also, to Professor Bryan Niblett for his review and helpful comments on my manuscript.

INTRODUCTION

Puzzle-solving has been a popular hobby for several centuries. Challenging and intellectually satisfying puzzles are usually the product of some considerable ingenuity. The work of the great puzzlists is as timeless as the great works of literature, and it is not meant as a criticism to concede that many of the best problems are somewhat artificial confections, improbable sets of circumstances depending on a certain suspension of disbelief on the part of the reader. Others are exercises in fairly elementary mathematics, the solutions depending on how much you remember from your schooldays.

Puzzle book writers are a prolific breed and produce a great many books, mostly borrowing freely and building on each other's works, modernizing and perfecting as they go, to intrigue new and old readers. Then there are the classics passed on from generation to generation, still handsome in their originality, but with their beauty slightly fading with the passage of time.

We also meet the endless "variations on a theme" which come under the heading: "If you've solved one, you've solved them all." Occasionally, but very rarely, a new idea is born which is pounced on eagerly by the puzzle fraternity.

At the same time, the physical world around us is a good provider of intriguing puzzles in its own right. This book is devoted to these conundrums, emanating from the works of nature which, in their perfection, are unmatched by anything the human mind can create.

Quite often we are unaware of the puzzles that confront us in everyday life and, even if we are, we rarely make the effort to try to understand the laws of nature which govern these phenomena. I suspect that some solutions will contradict your sense of logic, while others will make immediate sense, leaving you slightly frustrated for not having found them yourself. And that, I suggest, is the hallmark of the truly satisfying brain-teaser: whether man-made or existing in the real world, an infuriating blend of the obvious and the elusive.

ERWIN BRECHER

THE PUZZLES

Battleship in a Bathtub

Can you float a battleship in a bathtub? Specifically, the battleship at the top left below, which, floating on the high seas, has a mass of 30,000 tons.

Imagine slowly lowering the ship into a huge bathtub that is shaped like the ship but a little larger, containing a small volume of water. As the ship is lowered, the water is forced up the sides of the tub and over the rim until there is only a thin envelope of water between the tub and the ship's hull. Is it possible to float a 30,000-ton battleship in just a few hundred gallons of water? Keep in mind that a floating object displaces a volume of water equal to its weight and that 30,000 tons of water have a volume of several million gallons.

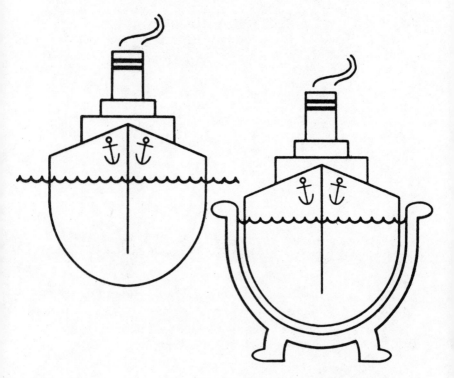

Solution on page 60.

The Leaning Tower of Dominoes

Stack dominoes on top of each other so that each domino projects as far as possible over the domino below without falling. If you stack two dominoes, the top one can overhang by as much as 50 percent of its own length. If you stack three dominoes, the greatest offset is obtained when the center of gravity of the top domino (A) is directly above the end of the domino below (B) and the combined center of gravity of these two dominoes is directly above the edge of the third domino (C). The top domino now overhangs the bottom one by 75 percent of its length.

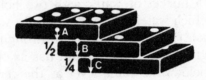

By continuing this pattern, is it possible to build a tower in which the top domino projects more than a full domino length beyond the bottom domino? If so, approximately how many dominoes do you think will be required to achieve this?

Solution on page 62.

The Light and the Shadow

A man is walking along a level road lit only by a solitary streetlight. He is moving at a constant speed and in a straight line as he passes the lamppost and leaves it behind, which makes his shadow lengthen. Does the top of the man's shadow move faster, slower, or at the same rate as the man? Prove your answer.

Solution on page 64.

The Bucket

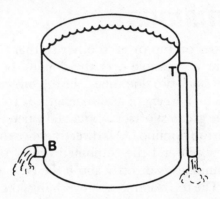

Consider a bucket of water with two holes of equal area through which water is discharged. The water can flow out through hole (B), at the bottom, or through the down-spout, which begins at the top (T) and has its opening the same distance below the water level as the center of hole (B). Ignoring any friction effects, out of which opening will the water flow faster, and why?

Solution on page 66.

Feathers and Gold I

Seriously, which weighs more, a pound of feathers or a pound of gold?

Solution on page 68.

SCI-BIT

Which phase of the moon is always in the opposite part of the sky to the sun?

The full-moon phase.

SCI-BIT
What was the first automatic safety device used in elevators?

The narrow shaft. The air-cushion effect produced is self-activating, thus automatic.

The Fast Car I

You are driving during rush-hour from Southtown to Northtown at a speed of 50 miles per hour. On your return journey there is little traffic, and you can maintain a steady speed of 100 miles per hour. What is your average speed for the whole journey?

Solution on page 72.

Balloon

An air-filled balloon is held underwater by a weight so that it is just on the verge of sinking; the top surface of the balloon just touches the waterline. If you push the balloon beneath the surface of the water (as shown), what will it do—bob back to the surface, remain at the level to which it is pushed, or sink to the bottom?

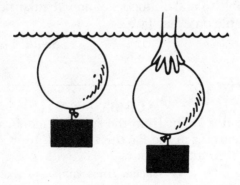

Solution on page 72.

Tumbling Polyhedrons

Each face of a polyhedron can serve as a base when the solid object is placed on a horizontal plane. With a regular polyhedron (all faces alike), the center of gravity is always directly above the center of a face; therefore it is stable on any face, as illustrated.

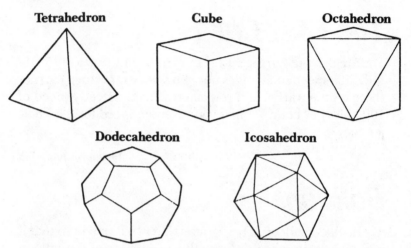

Tetrahedron **Cube** **Octahedron**

Dodecahedron **Icosahedron**

Irregular polyhedrons that are unstable on certain faces are easily constructed; that is, when placed on a plane with an unstable face at the bottom, the center of gravity is not directly above that base, and the polyhedron topples over.

Is it possible to make a model of an irregular polyhedron that is unstable on every face?

Solution on page 74.

Solution on page 74.

SCI-BIT

How long does light from the sun and moon, respectively, take to reach the Earth? 1 second, 10 seconds, 2 minutes, 8 minutes, 4 hours.

The moon, 1 second; the sun, 8 minutes.

Sunken Sub

The captain of a submarine tries at all costs to avoid letting his sub come to rest on a clay or sandy ocean bottom. He knows that if this happens, it can be fatal. Why?

Solution on page 76.

The Movement

What movement made by which muscle or limb or limbs is, without using any object such as a lever, the most powerful the human body is capable of?

Solution on page 78.

Bootstrap Elevator

Study the drawing carefully. Can the man lift both himself and the block off the ground?

Solution on page 80.

Iron Doughnut

A piece of solid iron in the shape of a doughnut is heated over a fire. As the iron expands, does the hole become larger or smaller or does it remain the same size?

Solution on page 82.

Balloon Behavior

A child sits in the back seat of the family car, holding a helium balloon on a string. All the windows are closed. As the car accelerates forward, does the balloon tilt forward, tilt backwards, or remain in the same place?

Solution on page 84.

The Unfriendly Liquids

There are two liquids that are available in most average homes which, when poured into an empty container, will never mix, flavor, or contaminate each other, and can be easily separated without heating or using any other substance. What are they?

Solution on page 63.

A Maritime Problem

Tourists on a cruise have experienced the less than enjoyable
sensation of pitching and rolling in a rough sea. Sailors must
either get used to it or look for another job.

On dry land, a similar sensation can be created by putting
oval wheels on a vehicle. Such wheels would make it pitch
backwards and forwards. Is it possible to make the vehicle
also "roll," and if so, how?

Solution on page 86.

Acceleration Test

We are told that the acceleration of an object in free-fall due
to gravity is the same for all bodies. This, however, is not
borne out by observation. In everyday life, objects do not fall
through the air at the same speed. A bullet and a piece of
paper dropped simultaneously will not hit the floor together.
Air resistance is supposed to be the cause, but can we be
certain? Short of trying it in a vacuum, which is not readily
available, can you think of a simple experiment that will
provide proof that acceleration is indeed the same for all
bodies?

Solution on page 62.

The Mirror Phenomenon

Why does a mirror reverse only the left and right sides but not up and down? After all, if we speak of a plain mirror, one with a surface that is perfectly smooth and flat, its left and right sides do not differ in any way from its top and bottom surfaces. Why then this persistent preference for changing left to right, while ignoring top and bottom? Can you explain this phenomenon?

While this question is indeed puzzling, we know intuitively that it could not be different. Remember, we are discussing the ordinary mirror, as it is quite possible to construct mirrors that will not reverse left and right, as well as others that will reverse top and bottom.

Solution on page 80.

Transparent Object

What transparent object, found in many households, becomes less transparent when wiped with a clean cloth?

Solution on page 68.

The Tall Mast

Joan and Jean are out sailing in a small boat when Jean remembers an important appointment. Joan suggests that they take a shortcut back to the dock to save time. On the way, they come to a footbridge and find that the boat's mast is slightly taller than the height of the bridge. How are they able to quickly pass under the bridge so that Jean can make her appointment? (They cannot lower the mast.)

Solution on page 65.

Corker

Take a glass that is partially filled with water and drop a cork into it. The cork won't float in the center, but keeps drifting to one side and attaching itself to the side of the glass.

   Without swirling the water, how can you make the cork float in the center?

Solution on page 66.

The Spectrum

One of the first art lessons learned at school is that the color green can be obtained by mixing yellow and blue paint. On the other hand, strangely enough, if you project yellow and blue light, through appropriate gelatin sheets, onto a screen, you obtain white light.
   Can you think of an explanation?

Solution on page 82.

SCI-BIT
What is the quantity of heat required to raise the temperature of 1 gram of water by 1 degree Centigrade?

One calorie.

The Brick and the Dinghy

An inflatable dinghy is floating in a swimming pool. Which will raise the water level higher: throwing a brick into the dinghy or throwing a brick into the water?

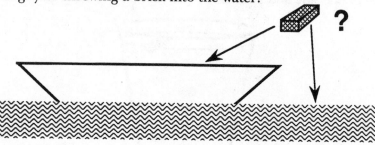

Solution on page 87.

The Glass Stopper

Individuals using perfume bottles with glass stoppers are often confronted with stoppers that cannot be easily removed without using force that could result in breakage. Placing it under a hot water tap won't work. Yet there is an easy method that works without fail. How would you solve this problem?

Solution on page 60.

The Fast Car II

Again, you are travelling from Southtown to Northtown. Midtown, as the name implies, lies exactly halfway between Southtown and Northtown. On your first leg of the trip, from Southtown to Midtown, you go a steady 50 miles per hour. At what speed would you have to drive the second leg from Midtown to Northtown to average 100 miles per hour for the whole one-way trip.

Solution on page 81.

The Bird Cage

There is a well-known, probably apocryphal story: A truck full of live poultry is stopped before a bridge on a country road. Its driver, beating the side of the truck with a stick, is asked what he is doing and he explains that his load is too heavy for the bridge so he is making the birds fly to lighten his load before proceeding.

This suggests the following puzzle. A cage with a bird in it, perched on a swing, weighs four pounds. Is the weight of the cage less if the bird is flying about the cage instead of sitting on the swing? Ignoring the fact that if left in an airtight box for long the bird would die, would the answer be different if an airtight box were substituted for the cage?

Solution on page 64.

And Yet It Moves

You have an academic background and are well read. You have studied the writings of Copernicus (1473–1543) including *De Revolutionibus Orbium Coelestium* (in translation), and of course you know all about Galileo Galilei (1564–1642) and his heliocentric view. But you are a skeptic and you wonder whether their theories can be accepted as the ultimate truth. It could just be that Ptolemy and his geocentric blueprint were right. After all, Galileo did recant, admittedly under pressure from the Church, and who knows if the story of his final words on his deathbed, "And yet it moves," is not merely an anecdote.

Now, suppose that money were no object and you had access to the most precise instruments. Could you think of at least three experiments that would prove to your satisfaction that Galileo was right, and it is the Earth that revolves and orbits the Sun?

Solution on page 81.

Hole Through the Earth

Assume that a hole is drilled from one point on the globe through the center of the Earth to the Antipodes, as illustrated, and a steel ball is dropped into the hole at point A.

Ignoring any external influences such as air resistance, friction and conditions of the Earth's core, answer the following questions:

1. As the ball is travelling from A to the Earth's center, does its velocity increase, decrease, or stay the same?

2. Will the ball weigh less or more when it reaches the center of the Earth?

3. Will the ball's mass change during its journey?

4. At what point will the ball be subject to zero gravity?

5. If the ball fell through such a cylindrical hole through the center of the moon, would the one-way journey take more or less time than it would on Earth?

Solution on page 88.

SCI-BIT

Which five planets can be seen with the naked eye?

Mercury, Venus, Mars, Jupiter, and Saturn.

Feathers and Gold II

Again, seriously, which weighs more, an ounce of feathers or an ounce of gold?

Solution on page 76.

The Half-Hidden Balance

The horizontal line is a weightless rod, balanced on a fulcrum at F, and of unknown length to the right of the fulcrum. The visible part is 1 foot long, and supports a 1-pound weight. Besides being weightless, the rod is able to support any weight, and, since it is in balance, there must be a weight to the right. The further the second weight is to the right, the less it can be. What are the lower and upper limits of the total possible force downwards at F?

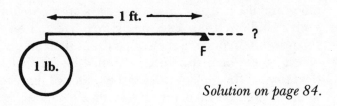

Solution on page 84.

Pendulum

If, in a vacuum, you hang two objects of the same weight by two strings of the same length they will swing in time with one another. If you increase the length of one string its weight will swing more slowly.

What will happen if instead you increase one weight?

Solution on page 68.

Salad Dressing

Jill and Joe went out for a picnic. "I hope you remembered to bring the oil and vinegar for the salad," said Joe.

"I certainly did," replied Jill, "and to save myself having to carry two bottles I put both the oil and vinegar in the same bottle."

"That wasn't very clever," said Joe, "because, as you very well know, I like a lot of oil and very little vinegar, but you like a lot of vinegar and hardly any oil."

Jill sighed, and then proceeded to pour, from the single bottle, exactly the right proportions of oil and vinegar that each of them wanted. How did she do it?

Solution on page 60.

The Asteroid

People have speculated that one day in the far future it may be possible to hollow out the interior of a large asteroid or moon and use it as a permanent space station. Assuming that such a hollowed-out asteroid is a perfect, non-rotating sphere with an outside shell of constant thickness, would an object inside, near the shell, be pulled by the shell's gravity field toward the shell or toward the center of the asteroid, or would it float permanently at the same location?

Solution on page 72.

SCI-BIT

Why can a needle float on water and small insects walk on water?

Because of the surface tension, an invisible skin or surface layer of water molecules.

Hourglass Puzzle

You can sometimes find an unusual toy in the shops: it is a glass cylinder full of water, with a sand-filled hourglass floating at the top.

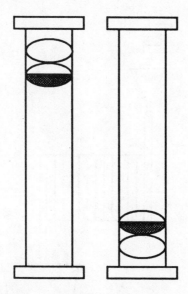

When the cylinder is turned upside down, as in the right-hand drawing, something rather strange happens. The hourglass stays at the bottom of the cylinder until a certain amount of sand has flowed from its upper compartment into its lower compartment, then it rises slowly to the top. Can you suggest a simple explanation for this phenomenon?

Solution on page 74.

Shut That Door

You are in a room filled with 100 percent methane gas. What would happen if you struck a match?

Solution on page 94.

The Garden Hose

The diagram shows a garden hose coiled around a reel about one foot in diameter, and placed on a bench.

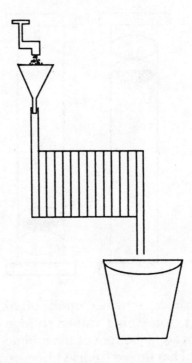

The lower end of the hose hangs down into a bucket. The upper end has been unwound so that it can be held several feet above the reel. When water is poured into the upper end using a funnel, one would expect that eventually it would run out at the lower end. Not so. Instead, as the water is poured into the funnel, it rises in the upper end of the hose until the funnel overflows. No water ever emerges from the lower end. Given that the hose is empty and has no kinks or obstructions, how do you explain this?

Solution on page 63.

Rotation Counter

If a bicycle with equal wheels has a rotation counter on each, why will the front one tend to give a higher reading?

Solution on page 88.

Soap Bubbles

Two soap bubbles (as shown below), unequal in size, are blown on the ends of a T-shaped tube.

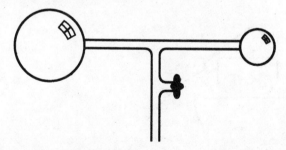

The blowing inlet is then closed, leaving an air passage connecting the two bubbles through the tube. What will happen? Will the smaller bubble expand at the expense of the larger one until they are both the same size? Or will something else happen? Or will nothing happen?

Solution on page 78.

Propelling the Boat

Assume that a rope has been tied to the stern of a small boat floating in still water. Is it possible for someone standing in the boat to propel it forward by jerking on the free end of the rope?

Now consider a space capsule drifting in interplanetary space. Could it be propelled by similar means?

Solution on page 64.

Two Cannonballs

Two identical cannons are aimed directly at each other, as in the illustration. The only difference is that the cannon on the left is at the top of a cliff and shooting downward, the one on the right is aiming up.

Neglecting air resistance, what will happen if the perfectly aligned charges are fired simultaneously and at the same speeds?

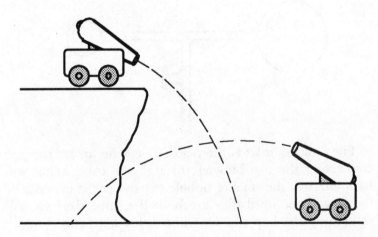

Solution on page 61.

Card Stack

It is impossible in reality, but assuming that one could tear a playing card in two, put the halves together and tear again (getting four), put them together again and tear, and so on *52 times*.

Do you think that the pile would be more or less than 10 miles high?

Solution on page 69.

Shooting Stars

On any clear night a shooting star can be seen in the sky about every ten minutes, on the average. It is commonly observed, however, that many more meteors are seen in the early-morning hours, between midnight and sunrise, than are seen in the evening between sunset and midnight. An astronomy student tells you that he thinks the reason for this is that artificial lights in homes and offices are usually turned off in the wee morning hours, which makes the whole sky more visible. What do you think of the student's reasoning?

Can you think of a better explanation for reports of morning meteors? Do Australians observe the same phenomenon, or is the situation reversed "down under"?

Solution on page 72.

Cool Off

Your air conditioner breaks down on the hottest day of the summer, leaving you sweltering in your one-room apartment. In order to get some relief, you leave your refrigerator door open. Will you cool the room that way?

Solution on page 90.

Center of Gravity

Suppose you are a manufacturer of cardboard boxes. Business is slow, and you are therefore more than happy to receive an inquiry for 50,000 boxes. The specification is somewhat unusual in as much as the shape of the box is to be irregular, as illustrated.

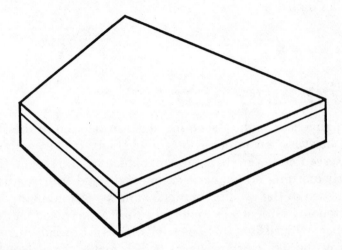

Furthermore, the customer, for his own reasons, wants you to mark the projection of the center of gravity of the box on the lid with a red dot. In other words, the empty box would balance on that point. You are asked to submit a sample with your quotation. How will you quickly find the center of gravity? (Note: this is not a geometry problem.)

Solution on page 80.

SCI-BIT
Why do some people put rice in their salt dispensers?

Rice absorbs and retains water more readily than salt, i.e., is more
hydroscopic, so keeps the salt dry.

The Camargue Horses

One of the great intellectual achievements of all time is Darwin's theory of natural selection. There cannot be many adults in the civilized world who have not heard of Charles Darwin (1809–1882) or his *On the Origin of Species* and the phrase "survival of the fittest." We understand why polar bears are white and why grasshoppers are green. Some of us even know that animals will survive if they are able to adapt to their environment. This adaptation is only possible by random genetic mutation or by favoring those members of a species whose characteristics make them better able to cope with the hazards of the environment.

A remarkable example of the workings of Darwin's theory is the story of the horses of the Camargue—the area of the Rhône delta in France. Originally, the horses roamed the area in multicolored herds, but eventually only the white horses survived. Can you explain this strange phenomenon?

Solution on page 88.

The Mystery of Weight

Many volumes could be written on the interconnected phenomena of weight, weightlessness and gravity. Letting our imaginations run wild, suppose we were to stand on a spring weighing machine at the equator and that, by some means, the Earth's rotation could be speeded up, carrying its atmosphere with it.

1. Would the weighing machine indicate a larger or smaller weight, or would there be no change?

2. If we speeded up the Earth's rotation further, would we reach a critical point, and if so, what would happen?

Solution on page 78.

The Helicopter

We are all more or less familiar with this heavier-than-air aircraft, which has one or more power-driven horizontal propellers that enable the craft to take off and land vertically, move in any direction, or hover stationary in the air. However, you may not have noticed the small rotor at the helicopter's tail. Is it important? What do you suppose is its purpose?

Solution on page 90.

Gravity Train

A German professor described a train that would travel from city to city, powered only by gravity: "Each railway is in a long tunnel, perfectly straight [that is, not following the Earth's curvature]; so of course the middle of it is nearer the center of the globe than the two ends; so every train runs halfway downhill, and that gives it force enough to run the other half uphill."

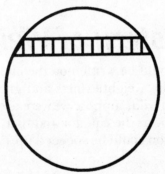

Ignoring friction and air resistance, as before, would such a gravity train work? And if so, about how long would it take to get from, say, Paris to Madrid? From Los Angeles to New York?

Solution on page 92.

Balance the Beaker

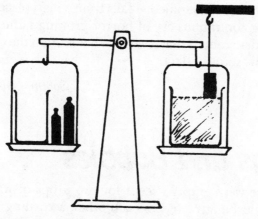

In the diagram above, there is a cylindrical iron bar, one square centimetre in cross-section, suspended vertically over a beaker. The beaker is two square centimetres in cross-section and partly filled with water. The bar just touches the surface of the water. The beaker is standing on the right-hand pan of a balance scale. The left-hand pan contains an empty beaker and sufficient gram weights to balance the scale. One cubic centimetre of water is added to the right-hand beaker. How much water must be added to the left-hand beaker to bring the scale back to balance?

Solution on page 84.

Death Fall

Frequently in movies someone falls from a high place and we hear a long continuous cry of horror growing fainter as the faller recedes from us downward. What is often the common error in this sound effect?

Solution on page 66.

Drops and Bubbles

If all space were empty except for two drops of water, the drops would be attracted to each other, according to Newton's Law of Gravity.

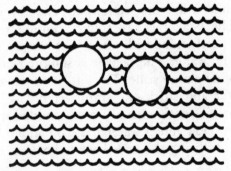

Now suppose all space were full of water except for two bubbles. Would the bubbles move apart, towards each other, or not at all?

Solution on page 89.

The Falling Elevator I

There must be few people who have not, at some time or other, travelled in a lift, or elevator. So probably everyone is acquainted with the feeling of extra personal weight when the car accelerates from rest to attain its steady upward velocity, together with a feeling of lightness when it decelerates to rest. Similar feelings are experienced in a descending elevator.

The behavior of a body in an accelerating or decelerating elevator is an important clue to solving the following:

A man entering an elevator holds a container filled with water. Inside the water there is a spring fastened to the bottom of the container at one end and to a cork at the other end, so that the cork is submerged just below the water surface.

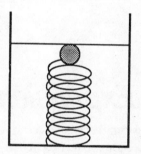

During the elevator's descent, the steel rope carrying the elevator breaks, and the elevator accelerates in free-fall. What happens to the container and its contents?

Solution on page 90.

Solution on page 90.

SCI-BIT

How long does light from the nearest star take to reach the Earth? 4 hours, 24 hours, 6 days, 4 years.

Four years, from Alpha Centuri.

The Falling Elevator II

Again in this puzzle the cable supporting an elevator breaks and the elevator falls. Assume this occurs without friction, so that the elevator falls with the acceleration of gravity. On the floor of the elevator is a drop of mercury. On the wall of the elevator a burning candle has been fixed. What happens to the mercury and the candle after the cable breaks?

Solution on page 79.

Bank Note

It is a felony to make a photograph of American currency, even if there is no intention of passing it, yet it is allowed on television. Why?

Solution on page 86.

Thermal Expansion

We know that most materials expand when heated. This property can be demonstrated easily in the interesting experiment below:

A flask, filled with water, is fitted with a rubber stopper, through which a tube is inserted into the flask and topped up with water until the water level rises a short distance up the tube to A–A.

The flask is then plunged into a container of hot water, and now the unexpected happens. The water level in the tube first falls to B–B before it rises steadily to C–C. Can you explain why?

Solution on page 91.

Rope and Pulley

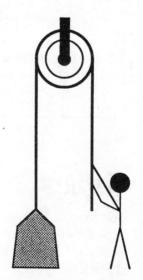

The diagram shows a rope passed over a frictionless pulley. To one end a weight is suspended, which exactly balances a man at the other end. What happens to the weight if the man attempts to climb the rope? Assume the rope to be weightless and the wheel to be frictionless.

Solution on page 70.

Solution on page 70.

Suspicion

Home alone, and gazing out of your window on a sunny afternoon, you suddenly notice a car parked in front of your local bank, with a driver at the wheel. Fumes from the exhaust tell you that the engine is running. Something seems amiss, but you don't want to alert the police. After all, it is only a hunch. That is when you realize that the license plate is in your line of vision. Unfortunately, you are shortsighted and left your only pair of glasses at the office. However you strain, you cannot see the numbers, and you don't own binoculars. Can you think of a way to enable you to read the license number from your window?

Solution on page 76.

Tables

You live in a hilly section of the suburbs and want to buy a new set of table and chairs for the garden. Would you be better off buying items of furniture that have four legs or three legs?

Solution on page 65.

The Tank

A closed glass tank is completely filled with water. A cork sphere is fixed to the bottom by a thread to prevent it from rising further. A steel sphere hangs from a thread fixed to the top of the tank. The lengths of the threads are such that the steel sphere is hanging just above the cork sphere.

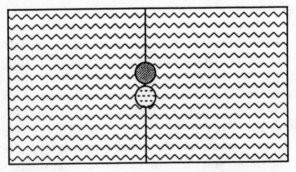

What happens if the tank is suddenly moved to the right?

Solution on page 73.

Two Bolts

Below are two identical bolts held together with their threads in mesh. While holding bolt *A* stationary, you swing bolt *B* around it counterclockwise, as shown.

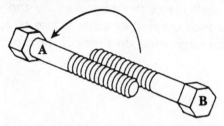

Will the bolt heads get nearer, move farther apart, or stay the same distance from each other? What will happen if you swing bolt *B* the opposite way—in a clockwise direction around *A*?

Solution on page 75.

Echo

When something makes a loud bang between us and the side of a house or a cliff, we hear it and its echo. What would we hear if the source of sound were touching the surface?

Solution on page 92.

Railways

The rails in the American and British railway systems have a length of 60 feet, while most of the railways in continental Europe use sections of 30 metres (98 feet 5 inches). If you examined them you would notice that all railway tracks have a small gap between adjoining sections. Why?

Solution on page 77.

The Smooth Table

I am at rest on a perfectly smooth, frictionless table, so that the only force acting on me is the upward pressure of the table due to my weight.

I can only shift my center of gravity up and down, and therefore I cannot crawl. The table is also so large that I cannot seize an edge. How do I get off the table?

Solution on page 89.

The Bottle and the Coin

A well-known parlor trick is shown below:

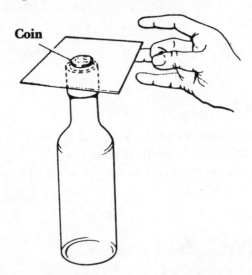

A small coin is put on a card (e.g., a visiting or business card) and placed over the mouth of a bottle. When the card is flicked away with the finger, the coin drops into the bottle.

Can you explain why this should be so?

Solution on page 85.

Accident in Space

You are the sole astronaut on a scientific mission in a spaceship. You rely on solar cell panels to provide energy, but discover that the system is malfunctioning. Repair should not be difficult, so you don your spacesuit and go outside.

As you hammer away, the umbilical cord becomes detached and the spacecraft moves slowly away from you. Not a pleasant situation, particularly as you are not equipped with directional jets, which would enable you to catch up with the ship.

You do have, however, a handgun with a 12-bullet magazine. If you were to fire the bullets in one-second intervals in the opposite direction, would you approach the spaceship? If so, would it be at uniform speed?

Suppose the craft is 3,300 feet (1,000 metres) away when you start firing. What precaution, if any, would you have to take?

Solution on page 74.

Sand on the Beach

Walk along a beach at low tide when the sand is firm and wet. At each step the sand immediately around your foot dries out and turns white. Why? The popular answer, that your weight "squeezes the water out," is incorrect: sand does not behave like a sponge. So what does cause the whitening?

Solution on page 86.

SCI-BIT

How many stars are visible to the naked eye on a clear night at any one moment? 1000–1200, 2000–2500, 5000–6000, over 10,000.

You can see about 2000–2500.

Deep-Sea Diving

After diving to a depth of about 180 metres with scuba equipment and remaining there only a few moments, a diver has to undergo nearly twelve hours of decompression in order to avoid a serious and possibly fatal attack of the bends. Can you explain what happens in rapid decompression?

Solution on page 69.

Table and Box

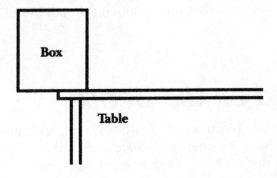

The box illustrated is resting on a table, projecting slightly more than half its width over the edge. It can remain in this position owing to the weight of something inside it. When the table is bumped vertically, the box falls. What does the box contain? Its contents can be found in most households.

Imagine a similar box, but projecting less than halfway over the edge of the table. In this case it can be made to fall without bumping, touching, or applying any pressure to it or to the table, directly or indirectly. How? What is in this box?

Solution on page 91.

Quito

Suspend your disbelief now for this most intriguing puzzle. Approximately how much would a man living in Quito, Ecuador, weigh if he were 44,000 miles tall? Would it be more or less than a thousand tons? (This needs a great deal of thought, not just an inspired guess.)

Solution on page 82.

Athletics

The moon's gravity is about one-sixth of that of Earth. If you weigh 120 pounds on your bathroom scale, you'll weigh a mere 20 on the moon. An object tossed straight up on the moon will go six times as high as it would on Earth if thrown with the same force.

The first Lunar Olympics will be held in an enclosed arena, warmed and air-filled, so that athletes won't be encumbered by space suits. If a high-jumper can clear a six-foot bar on Earth, to what height will he or she be able to jump on the moon? The answer usually given—36 feet—is incorrect.

Solution on page 87.

SCI-BIT

What are these sciences? (a) genetics; (b) cybernetics; (c) kinetics.

(a) heredity and variation in animals and plants; (b) control and communication mechanisms in animals and machines; (c) relations between the motions of bodies.

Big into Small

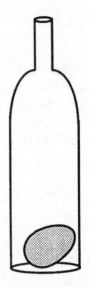

To put big objects into bottles has become something of an art. It seems as impossible as passing a camel through the eye of a needle, but these objects include such items as pears, arrows, and ships. Various techniques are used that might differ from object to object.

Our puzzle deals with a classic of this type, namely a hard-boiled egg into a milk bottle.

The method is pretty well known. First you drop a piece of paper and a lighted match into the bottle, then place a peeled egg upright over the neck. The oxygen consumed by the fire creates a partial vacuum and the atmospheric pressure will push the egg into the bottle. After you wash the ashes and dead match out, you can take a bow.

But, can you do the same with an *unpeeled* egg?

Solution on page 85.

SCI-BIT

Is water a good or poor conductor of: (a) electricity; (b) heat; (c) sound?

(a) good; (b) poor; (c) good.

Gravity

About how much, do you think, is the moon affected by the gravitational pull of the sun, as compared with that of the Earth? The moon stays with us and doesn't go flying off toward the sun, so the Earth's attraction is greater, right? Wrong. The sun pulls more than twice as hard as the Earth. So why hasn't the sun stolen our moon away?

Solution on page 81.

Earth and Moon

The motion of the Earth and moon causes the tides, and the ceaseless sloshing of the tides is having an effect on the motion of the Earth. What is happening?

Solution on page 89.

Blowing Out the Candle

Take a lighted candle and a sheet of paper, which you roll into a hollow cone, as shown.

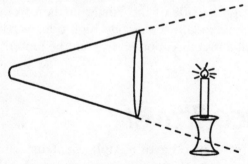

If you try to blow the candle out by blowing through the small end, you will fail. How can you be more successful?

Solution on page 75.

Shoot the Moon

If you want to take a picture of the full moon and get the largest possible image on film, should you shoot it when it is directly overhead (and therefore at its closest point to your position on Earth), or when it is down hear the horizon? Almost everyone says that the moon is largest near the horizon. Is this an atmospheric effect or a psychological one? Does it show up in photographs?

Solution on page 63.

The Tides

The ocean tides are caused by the gravitational pull of the moon and sun, mostly of the moon. But why are there two high tides a day? When there's a high tide on the side of the Earth closest to the moon, there is simultaneously a high tide on the opposite side of the Earth, farthest from the moon. How do you explain the second high tide? What makes the water bulge in a direction *away* from the moon?

Solution on page 65.

Space Travel

The nearest star to Earth is Alpha Centauri, which is about 4.3 light-years away. Assuming we can travel at the speed of light, what determines the amount of time our intrepid voyagers would need to reach the star?

Solution on page 69.

Free-Fall

A cork is suspended in the middle of a bucket of water held down by a fixture to prevent it rising to the top, as shown in the illustration.

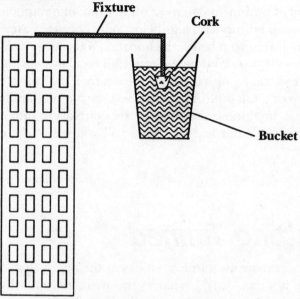

The bucket is then dropped with the fixture remaining on the roof, so that the cork is free to move. Where will the cork be just before the bucket hits the ground?

Solution on page 86.

The Rainbow

The rainbow is one of nature's most uplifting spectacles. Occurring as it does when sunshine meets suspended water droplets, as after a summer shower, the rainbow is a universal symbol of optimism. We most often think of a rainbow as an arc, round on top with legs below, only because water droplets are rarely seen below the horizon. You can see full-circle rainbows if you stand near a waterfall or a lawn sprinkler, or if your vantage point is "above the weather," as from a cliff or the top of a tall building. When you see a more familiar arc rainbow, imagine extending the arc's curvature into a complete circle. What feature, then, will you see at the circle's center?

Solution on page 73.

Getting Tanned

You'll get more suntan from a day at the beach than from a day in the backyard. What is the main reason why your ultraviolet exposure is more at the shore?

Solution on page 77.

Climbing the Mountain

We were all climbing the local mountain, and as we got higher it seemed to get colder and colder. John said it was due to our greater exposure to the cold winds. Bob said it was something to do with the more rarefied atmosphere. Tom said it was just imagination and as we were nearer the sun it must really be warmer. Bill said we were farther from the center of the Earth, which was known to be hot. What do you say?

Solution on page 75.

Saving Syracuse

According to a famous legend, when Roman ships attacked Syracuse in 214 B.C., Archimedes saved the day with mirrors. He positioned soldiers on the shore, each with a large mirror. At a signal, they burned the Roman fleet by reflecting sunlight into the ships.

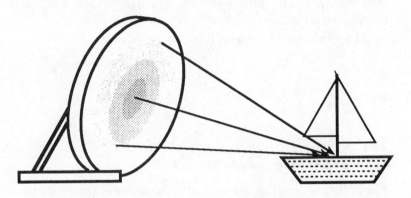

Fiction aside, is there any practical validity to Archimedes' tale? Whether or not he really tried to defend Syracuse with mirrors, we probably will never know, but could his trick have worked?

Solution on page 79.

The Flawed Sense

The five faculties through which we perceive the world—sight, smell, taste, hearing, touch—are, together with the brain, a most complex communication network, unsurpassed by any computer system created by man. However, one of our senses is flawed, and it is that flaw which enables us to enjoy a leisure activity which dominates our daily life. What is it?

Solution on page 78.

The Atmosphere

One of the properties of the atmosphere is air pressure, which shapes the physical world in which we live. Because we do not feel it, we don't give it a second thought, and we are blissfully unaware of the profound effect this phenomenon has on our existence. We are not even conscious of the load exerted on us by atmospheric pressure, which for the average-sized man can exceed twelve tons.

Why does this enormous load not crush us?

Solution on page 70.

Skipping Stones

The ability to skip flat stones across the water is a question of skill. It is difficult to measure the path of a stone across water, but if you skip a stone on the wet sand at the water's edge, it will leave marks tracing its path. The flight is surprisingly complex. Long hops of several feet alternate with short hops of just a few inches, and zigs to the left alternate with zags to the right. A right-handed throw, with the proper grip, spins clockwise and strikes the sand first with its trailing edge.

Proper grip

Will the first hop be short or long? To the left or the right?

Solution on page 85.

The Spinning Top

The most famous spinning toy of all is the top, and the most perplexing version of it is the Tippe Top (below).

Initial orientation

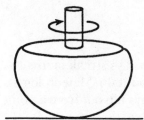

Final orientation

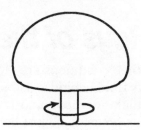

Its behavior is so paradoxical that whole treatises about it have appeared in *Physica*, *The American Journal of Physics*, and other scientific periodicals. Physicists are still not in agreement as to how it works.

The Tippe Top has a spherical bottom and a stem on its top. The surprising thing is that when you spin it on its spherical surface, it stays on this side for only a few seconds, then turns upside down and spins on its stem. This raises the top's center of gravity, which would seem to violate the principle of the conservation of energy.

You can demonstrate this paradox to yourself with any Tippe Top or with any typical high-school or college-class ring that has a smooth stone. Spin the ring on its stone, and in a few seconds the ring will invert itself and spin with the stone facing upwards. A hard-boiled egg that is spun flat will rise to spin on one end. (Spinning an egg is a good way to tell whether it is hard-boiled; a fresh egg won't spin because the liquid inside sloshes around too much.)

Do you have an explanation?

Solution on page 87.

Rays of the Sun

When conditions are right, the sun's rays streak across the sky from behind a distant cloud or mountain. Meteorologists call these crepuscular rays. They always fan out from a point that seems to be just behind the obstruction.

But wait, aren't the sun's rays supposed to be parallel when they reach Earth? How does a cloud or mountain cause the rays to diverge that way?

Solution on page 89.

Vacation on the Moon

Edgar D. Twitchell, a New Jersey plumber, was on his way to the moon for a three-week holiday. The rocket ship was too small to generate artificial gravity by spinning, so Twitchell had the strange sensation of feeling his weight steadily diminish as the ship sped towards its destination. When it reached the spot where Earth's stronger gravity field was exactly balanced by the moon's weaker field, zero gravity prevailed inside the ship. All passengers were kept fastened to their seats, but Twitchell enjoyed the floating feeling nonetheless as he twiddled his thumbs and contentedly puffed a cigar.

Many hours later the ship slowly settled next to one of the huge domes that house the U.S. moon colony, its descent cushioned by rocket brakes. Through the thick glass window by his seat Twitchell caught his first glimpse of the spectacular lunar landscape. Several large seagulls, with tiny oxygen tanks strapped to their backs, were flying near the dome. Above the dome an American flag fluttered in the breeze.

Although it was daylight, the sky was inky black and splattered with twinkling stars. Low on the horizon a rising "New Earth" showed a thin bluish crescent of light with several faint stars shining between the crescent's arms. As Twitchell later learned, the moon makes one rotation during each revolution around the Earth. Because a rotation takes about twenty-eight days, it takes the Earth about fourteen days to rise and set on the moon.

On the sixth day of his vacation, Twitchell was allowed to put on a space suit and hike around the crater in which the dome had been built. After bounding along for a while he came upon a group of children, in pink space suits, playing with boomerangs. One girl tossed a boomerang that made a wide circle and Twitchell had to duck as it whirled past his helmet. Behind him he heard it thud against a large boulder. He turned to look, but the curved stick had fallen into the rock's ebony shadow where it instantly seemed to vanish. Since there is no atmospheric scattering of light on the moon, objects cannot be seen in shadows without a flashlight.

The sun was low in the sky when Twitchell began his walk. Now it was sinking out of sight. The "terminator," that sharp line separating the lunar day from night, was gliding across the grey terrain toward the brightly lit dome at a speed of about 40 miles an hour—much too fast for Twitchell to keep up with it by vigorous hopping. Overhead, a meteor left a fiery trail as it fell to the moon's surface.

Twitchell was so exhausted when he returned to his quarters that he fell asleep on his bed, fully clothed, and did not awake until the rising sun flooded his room with brilliant sunlight.

How many scientific mistakes can you find in the above narrative?

Solution on page 82.

The Space Station

Weightlessness in a space vehicle is highly inconvenient to any astronaut in many ways. For example, he cannot pour liquid into a cup, neither can he drink from it: controlled movement is possible only by the use of handrails and so on.

It has been suggested that the space stations of the future for use of manned observatories or as staging posts for space exploration might be built in the form of huge wheels with hollow rims, as illustrated. These would be set in rotation so that the outer rim, which acts as the floor, would apply a radial centripetal force to the occupants or any objects inside to keep them moving in a circle.

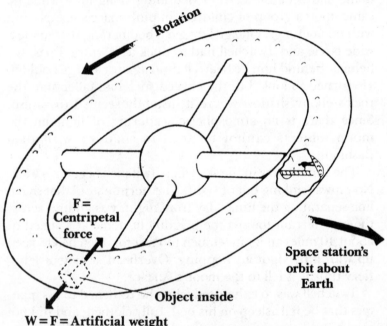

Rotation

F =
**Centripetal
force**

Object inside

W = F = Artificial weight

**Space station's
orbit about
Earth**

Artificial Weight in a Space Station

The equal and opposite reaction to this centripetal force which any person or object exerts on the floor would act as an artificial weight, thus allowing eating, drinking, and working in comparative comfort. The quantum of this weight could be made equal to or less than the normal Earth weight simply by adjustment to the speed of rotation.

1. What would happen to an astronaut's weight if he were to walk round the space station in the direction of its rotation and then turn round and walk in the opposite direction?

2. Now suppose you are in a small windowless room in the same space station after suffering a bout of amnesia. In other words, you do not remember that you are in space. The speed of rotation around the hub produces a simulated gravity of one g. Inside your room everything seems "normal"—gravity seems to be operating on you exactly as it would on Earth. In fact, as far as your senses tell you, you are on Earth.

In your pocket you have a magnet, a piece of string, some coins, a pencil and a steel paper clip. Suddenly you have some doubt as to where you are. Is there a simple test you could do in your room, using one or more of the objects in your pocket, which would confirm that you are on a spinning space station and not on Earth?

Solution on page 91.

The Wrong Way

There is a solid, physical part of a travelling railway train which moves in the opposite direction to the rest of the train. It is a part of the train, not of its contents. The part is firmly attached to the train and remains with it all the time, yet the part is going backwards when all the rest of the train is going forward.

Actually, there are many such parts on every travelling train. What is the so-called "part"?

Solution on page 61.

Field of Play

The World Cup playing field was painstakingly constructed to be a perfect plane. During the national anthem the 22 players, the referee, and the 4 linesmen were standing at attention. However, it was unlikely that any one of them, but at best only one, actually stood upright. Explain.

Solution on page 73.

Evolution

The theory of evolution by Charles Darwin (1809–1882) must be considered one of the great intellectual achievements of all time. His principle of natural selection, explaining the process of adaptation that provides species with optimum survival characteristics, has many applications in the world around us. Here is one:

Ornithologists have found a convincing explanation, based on Darwin's principle, why birds' eggs are generally narrower at one end than at the other. Can you?

Solution on page 81.

Litmus Test

Litmus paper is widely used for laboratory purposes. It consists of an absorbent paper impregnated with a dye, and is the oldest form of acid–base indicator. Dipped into an acid solution it will turn red, and in a base solution it will turn blue.

If no litmus paper is available, do you know a natural product which will serve the same purpose?

Solution on page 75.

The Electric Fan

Why is it an advantage to have a constantly running electric fan in a room heated by a wood or kerosene stove, but not in one heated by a big fireplace?

Solution on page 63.

The Brick Bridge

A farmer wants to build a bridge made of bricks, over a small river. His son, a student architect, suggests two designs as shown:

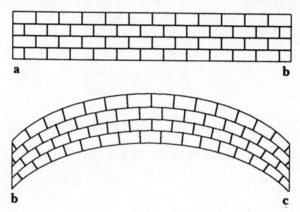

1. Which design can carry heavier traffic?

2. Which will need more building material?

Solution on page 65.

The Mercury Experiment

Fill a glass test tube, open at one end, with mercury. Close the open end with a finger and then invert the tube into a vessel filled with mercury, as shown in the illustration below.

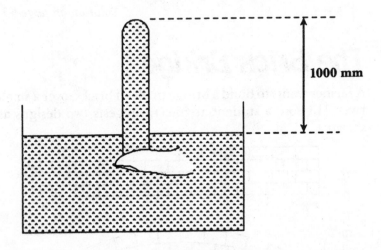

1000 mm

1. What happens when the finger is removed?

2. What practical use is made of this experiment?

Solution on page 67.

Whiskey and Soda

You have a quart of soda water and a quart of whiskey. Take one tablespoonful of soda, transfer it to the whiskey, and mix thoroughly. Then take a tablespoonful of this mixture and pour it back into the soda. Is the amount of soda in the whiskey greater than, less than, or equal to the amount of whiskey in the soda?

Solution on page 69.

The Archimedes Puzzle

Legend has it that Archimedes was asked to verify whether a crown was made of pure gold, as was alleged, or whether it had been adulterated with silver.

Suppose you know the specific weight of gold, how would you go about the problem, short of melting down the crown? Remember that gold is heavier than silver per unit volume (specific weight).

Solution on page 70.

Three-Dimensional Vision

There have been a number of attempts to create a three-dimensional effect from two-dimensional images. One of the earliest was a device called a stereoscope, through which one looked at specially printed cards. Much later, "3-D" came to the cinema screen. Through red- and green-lensed spectacles passed out upon entering the theatre, cinema-goers enjoyed an effect that was said to be so realistic that a train speeding out of the screen towards the audience created panic. This experiment was soon abandoned as the 3-D glasses needed found little favor with movie audiences.

Since then, more sophisticated techniques have been devised, leading to specially designed patterns (stereograms), usually printed in color, that are being marketed under various trade names. Once the viewer has mastered the technique, the effect is spectacular.

It is, of course, one thing to enjoy the images, but quite another to understand how it works. Can you find an explanation for the 3-D effect of the patterns, and prove that your theory is correct?

Solution on page 92.

The Mix

A man is heating liquid A which will be mixed with liquid B, now cold. If he continues heating A, the resultant mixture of A and B will be less hot than if he were to mix it now. Why?

Solution on page 85.

The Wheelbarrow

Which is easier: pushing a wheelbarrow or pulling it?

Solution on page 84.

SCI-BIT

Which word describes: (a) liquid that does not flow freely; (b) metal that can be shaped by hammering; (c) metal that can be drawn out into wire or thread?

(a) viscous; (b) malleable; (c) ductile.

58

THE SOLUTIONS

Battleship in a Bathtub

Strange as it seems, it is perfectly possible to float a battleship in a bathtub. As long as there is enough water to surround the ship completely, it will float. The ship's hull can't "tell" whether it is surrounded by an expanse of ocean or by a mere fraction of an inch of water. The water pressure on the hull is the same in either case. Likewise, the hydrostatic pressure of the ship is independent of the amount of water that is below or to the side of the ship.

Many people find this hard to believe because they confuse the amount of water "displaced" with the amount necessary to float the ship. Look at it this way: Suppose the ship weighs 30,000 tons. If the tub is filled to the rim with several million gallons of water and you lower the ship in, the amount of water that will spill over the sides will, indeed, weigh 30,000 tons. But that's water over the rim. The amount of water left inside, floating the ship, could be a great deal less. If the "fit" between the ship's hull and the tub's wall is tight enough, the water could be squeezed into a thin envelope only an inch thick or less, completely surrounding the ship. This principle is put into practice at Mount Palomar Observatory, where the giant 530-ton horseshoe telescope actually floats in a basin on a thin cushion of oil.

The Glass Stopper

You will likely need someone's help with this: another pair of hands to hold the bottle firmly while you wrap a narrow strip of cloth around the bottle's neck. Then, by pulling the cloth rapidly to and fro, you can cause friction between the strip and the glass that generates sufficient heat to expand only the bottle's neck so that the stopper can be withdrawn easily.

Heating otherwise does not do the trick, as bottle and stopper would expand uniformly.

Salad Dressing

Since oil floats on vinegar, to pour mostly oil Jill had only to uncork and tip the bottle, then cork it, turn it upside down and loosen the cork just sufficiently to allow the desired amount of vinegar to dribble out.

Two Cannonballs

Surprisingly, no matter how far apart the cannons are, or at what angle they aim at each other, the balls will always collide in flight. If there were no gravity, the missiles would meet midway between the two cannons. With gravity, the balls fall below this point by equal distances; so they will still collide in mid-air. This is a special case of the fixed-point theorem in topology.

The Wrong Way

The many parts of the train that are going the wrong way are the tips of the flanges of each of the wheels.

Let XX be the position at rest, YY the position after the wheel has moved through 60 degrees, and DD the destination. The double line represents the rail.

After one-sixth of a revolution (60 degrees), A becomes AA, and B becomes BB. It is obvious that the distance AA to DD is shorter than A to DD, while BB to DD is longer than B to DD. Hence B has moved backwards.

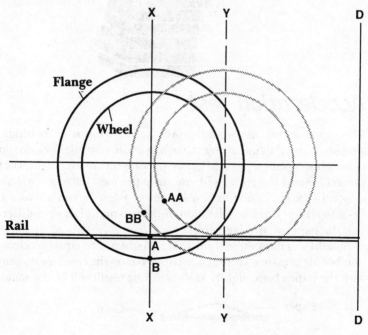

The Leaning Tower of Dominoes

Yes, it's possible. In fact, the top domino can be made to overhang as far as you want beyond the bottom domino without toppling the structure. Success in building such a tower depends on working down from the top; you have to know how many dominoes will be in the final tower before you start stacking. As shown, the top domino extends half a domino length. The second projects ¼; the third, ⅙; the fourth, ⅛; and so on.

After only five dominoes are placed, the top domino (A) is already offset by more than its own length over the fifth domino. With an unlimited supply of dominoes, the offset is the limit of ½ + ¼ + ⅙ + ⅛ + ⅒ + ... It is a series that can grow as large as any finite number. On the fiftieth term, the overhang is about two and one-quarter lengths. For three lengths you'll need 227 dominoes, and for 10 lengths you need 1.5×10^{44}.

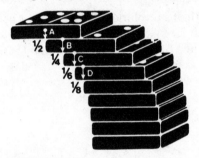

Acceleration Test

The experiment that would prove the doctrine is surprisingly simple. Take a flat, metal object, such as a coin or medallion, on top of which you place a piece of paper, making certain that it does not project beyond the rim. Hold and drop the metal object horizontally, with the paper on top. You will observe that the two will hit the floor together. If the acceleration of the two materials were different, the paper would lag behind.

Should you happen to think that it might not be equal acceleration but air pressure keeping the paper flat on the coin, try it again with the paper bent slightly as shown. The result will be the same.

Paper ─────────────────────── Coin

The Garden Hose

When water is first poured in, it flows over the first winding of the hose and forms an air trap at the bottom. This trapped air prevents any more water from entering the first loop of the hose.

If the funnel end of the empty hose is high enough, then water poured in will be forced over more than one winding to form a series of "heads" in each coil. The maximum height of each head is approximately equal to the diameter of the coil. The diameter multiplied by the number of coils gives the approximate height the water column at the top end has to be in order to force water out at the lower end.

The Unfriendly Liquids

The liquids are mercury (from a thermometer) and water. The water stays on top and can be easily decanted. Any traces left on the surface of the mercury can be removed, if necessary, by blotting. Oils won't work, as they would always leave a thin but unremovable film on both water and mercury.

In case the reader would argue that blotting paper is "another substance," one can do without it by simply waiting a short while until the thin film of water evaporates.

Shoot the Moon

Photographs show that the moon is the same size in both positions. The illusion is universal—it is apparent even in a planetarium, but there is still no fully accepted explanation for it. Ptolemy argued that the horizon moon appears larger because we can compare it with distant trees and buildings. This theory is still the most widely accepted one, but it doesn't explain why sailors see the moon illusion just as vividly at sea.

The Electric Fan

It circulates the hot air that stoves give off. Fireplaces (non-smoking ones) only give off radiant heat, which is unblowable.

The Light and the Shadow

The top of the shadow moves faster than the man.

As proof, let A be the position of the man at one point in time, and B be the man's position after he has walked 20 metres. Let AA and BB be the top of the shadows in the two positions. The distance AA–BB is clearly greater than A–B so, consequently, the top of the shadow must have moved faster.

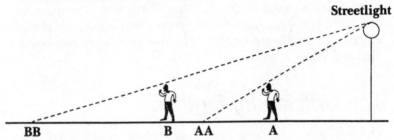

The Bird Cage

If the bird is in a completely airtight box, the weight of the box and the bird will be the same whether the bird is flying or perching. If the bird is flying, its weight is borne by the air pressure on its wings; but this pressure is then transmitted by the air to the floor of the box. If the bird is flying in an open cage, part of the increase in pressure on the air is transmitted to the floor of the cage, but part is transmitted to the atmosphere outside the cage. Hence the cage with the bird will be lighter if the bird is flying.

Propelling the Boat

By jerking on a rope attached to its stern, a small boat can indeed be moved forward in still water, and speeds of several miles per hour can be achieved. As the person's body moves toward the bow, the friction between the boat and the water prevents any significant backwards movement of the boat, but the inertial force of the jerk is strong enough to overcome the resistance of the water and transmit a forward impulse to the boat.

The space capsule, in the absence of friction, cannot be propelled in the same way.

The Tall Mast

They open the stopcocks and let some water into the boat. This makes it ride lower in the water so that the mast clears the bridge.

Tables

A table with three legs is more stable, as three points always lie in a plane.

The Tides

It is convenient to speak of the moon going around the Earth, but it is more accurate to say that they form a two-body system that revolves around its center. This common center of gravity is actually inside the Earth because of the Earth's greater mass. As the Earth swings around this center, centrifugal force causes the ocean waters to flow away from the center, producing two outward bulges on opposite sides of the globe.

The Brick Bridge

1. The *ab* design can accommodate a heavier burden.

2. The amount of material needed is the same for both designs.

Bridge *ab* is the same length as bridge *bc*. If bridge *bc* were sliced along the dotted line *1* and the upper section moved down to dotted line *2* we would have a brick bridge of the same dimensions as bridge *ab*. This clearly proves that both bridges contain the same amount of material.

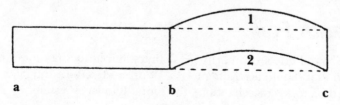

The Bucket

The speed of the flow depends on how far below the water's surface the hole is located—or its depth. The depth is the same for both outlets, so water will come out of both holes at the same speed. There's a simple intuitive proof of this problem. Imagine that the two outlets were on the same side of the bucket and joined, as shown below.

If water flowed faster through either opening, it would overwhelm the water in the other opening, pushing it back into the bucket and creating a perpetual-motion machine.

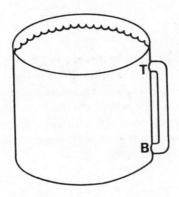

Corker

Pour a bit more water into the glass, not only until it is full, but until the surface of the water is slightly convex above the rim (this is possible because of surface tension). The cork will naturally float to the highest point of the water dome—in the center—and remain there.

Death Fall

The sound effect ordinarily gets fainter, but it does not lower in pitch as it should according to the Doppler effect—as the faller's speed increases (acceleration), the sound frequency lessens.

The Mercury Experiment

1. The mercury level will drop to a point, A, such that the pressure of the atmosphere of about 14.7 pounds per square inch exerted on surface S is balanced by the weight of the mercury column A over S (approximately 30 inches, fluctuating with air pressure).

2. This phenomenon was first used by Evangelista Torricelli in 1643 to construct a barometer. Theoretically any liquid could be used for this purpose, but since mercury is about 13.6 times heavier than an equal volume of water, it follows that the water column would have to be about 13.6 times as high as the mercury column, that is, 34 feet of water corresponding to 30 inches of mercury.

Torricelli explained that the column of mercury was supported in the tube by the atmospheric pressure acting on the surface of the mercury in the dish, and pointed out that small changes in the height of the column, which are noticed from day to day, are due to variations in the atmospheric pressure. The space above the mercury in the tube is called a Torricellian vacuum; it contains a little mercury vapor, and in this respect differs from a true vacuum.

Torricelli died a few years after the barometer experiment had been performed, and did not live to see his explanation of it, in terms of atmospheric pressure, generally accepted among scientists.

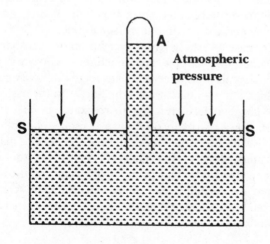

Feathers and Gold I

A pound of feathers weighs more. This is because feathers are weighed using the Avoirdupois system, whereas gold (as are silver and drugs) is weighed using the Troy system. A pound on the Avoirdupois scale is greater, therefore, than a Troy pound.

Transparent Object

Wet ground glass. The clean cloth must be dry. The commonest example in a kitchen might be a Pyrex baking dish that has become worn or scratched on the bottom so as to be merely translucent, not transparent. When wet it is clear.

Pendulum

No change. Everything falls at the same speed, and gravity is the sole cause of the pendulum's downward motion. This is an important phenomenon in physics.

Galileo Galilei was the first scientist to recognize this principle. He is particularly remembered for the work he did on the acceleration of falling bodies. The Greek philosopher Aristotle had taught that the speed with which a body fell to the ground depended on its mass. This wrongful assertion had been accepted for centuries, but legend has it that Galileo tested the matter by a very simple experiment. He ascended the leaning tower of Pisa and, from the top, simultaneously released three iron balls of different masses. They all reached the ground at the same time.

It is very easy to draw a wrong conclusion from casual observation. A feather, for example, falls much more slowly than a stone. The fact is that air resistance as well as force of gravity govern the rate at which a body falls towards the Earth. In the case of a light feather of large surface area, the air resistance is very great compared to the force of gravity on it. If air resistance is eliminated the feather falls with the same acceleration as the stone. This was first demonstrated, shortly after Galileo's death, by Robert Boyle. Using his newly invented air pump, Boyle removed the air from a tall glass jar containing a lead bullet and a feather. When the jar was inverted both bullet and feather reached the bottom of the jar simultaneously.

Card Stack

Higher than the sun is from the Earth, so more than 93 million miles.

Deep-Sea Diving

On rapid decompression the nitrogen (and oxygen), which under pressure dissolves in the blood, is released by the blood in the form of bubbles that create stoppages in the joints, lungs, spinal cord, and heart (embolism). Controlled decompression of 20 minutes for each atmosphere of pressure (about 15 pounds per square inch) allows the dissolved nitrogen to come out of solution slowly enough to be removed by the lungs without the formation of bubbles.

Space Travel

It would seem that we can get there in a minimum of 4.3 years, but we must instead determine the maximum acceleration and deceleration a human body can stand to reach and return from a speed which has been approximated at 186,000 miles per second.

Whiskey and Soda

There is the same amount of whiskey in the soda as there is soda in the whiskey. This classic puzzle can be attacked with algebra, fractions, and formulas, or with intuition.

Both containers are left with a quart of liquid after the transfer, the same volume they had to start with. The amount of whiskey missing from its bottle is exactly matched by enough soda to bring the bottle's volume back up to a quart. Likewise, the amount of soda removed from its carton, whatever it may be, is now taken up by an equal volume of whiskey.

This is true of whiskey and soda, but not with all pairs of liquids, such as if a spoonful of one liquid plus a spoonful of the other becomes less (or more) than two spoonsful of mixture due to chemical reaction.

The Atmosphere

We do not feel the enormous load exerted on us because our blood pressure balances the atmospheric pressure.

An historic experiment, the Magdeburg Hemispheres, performed in 1654 by the German scientist Otto von Guericke before the Imperial Court, demonstrated the enormous pressure exerted by the air around us.

Two hollow metal hemispheres of about 30 inches in diameter were provided with an air-tight leather ring separating their rims. Air was pumped out through a tap which was then closed, creating a vacuum inside the sphere as illustrated:

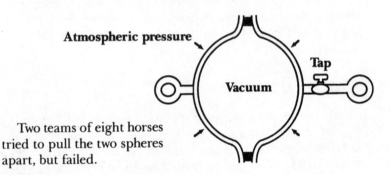

Two teams of eight horses tried to pull the two spheres apart, but failed.

Rope and Pulley

The answer is that, regardless of how the man climbs—fast, slow, or by jumps—the man and weight always remain opposite. The man cannot get above or below the weight, even by letting go of the rope, dropping, and grabbing the rope again.

The Archimedes Puzzle

Specific weight is the ratio of the weight of a solid to the weight (mass) of an equal volume of water (at 4 degrees Centigrade).

Archimedes could weigh the crown, but his problem was to find the volume of an object as complicated in design as a crown.

In a flash (Eureka! I have found it) he realized that if he

immersed the crown in water the volume of water displaced was equal to the volume of the immersed object. Comparing the weight of the crown with the weight of the displaced water gave him the specific weight which, if less than that of gold, would indicate that the metal had been debased.

While the solution of this problem is a considerable intellectual feat on the part of Archimedes, it is dwarfed by his most important discovery which is now called *Archimedes' principle*:

When a body is wholly or partially immersed in a liquid it experiences an upthrust equal to the weight of the liquid displaced.

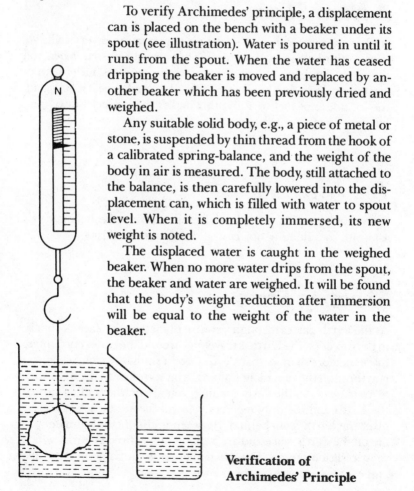

To verify Archimedes' principle, a displacement can is placed on the bench with a beaker under its spout (see illustration). Water is poured in until it runs from the spout. When the water has ceased dripping the beaker is moved and replaced by another beaker which has been previously dried and weighed.

Any suitable solid body, e.g., a piece of metal or stone, is suspended by thin thread from the hook of a calibrated spring-balance, and the weight of the body in air is measured. The body, still attached to the balance, is then carefully lowered into the displacement can, which is filled with water to spout level. When it is completely immersed, its new weight is noted.

The displaced water is caught in the weighed beaker. When no more water drips from the spout, the beaker and water are weighed. It will be found that the body's weight reduction after immersion will be equal to the weight of the water in the beaker.

**Verification of
Archimedes' Principle**

The Fast Car I

Your average speed is 66.67 miles per hour, not 75 miles per hour. You cannot average the two speeds because you spend twice as long driving at 50 miles per hour as you do at 100 miles per hour on your return journey.

The point is that speed equals distance divided by time, and it is this formula which must be used when calculating an average.

A Hard Skate

It's harder to ice-skate when the air temperature is very cold. We tend to think of ice as inherently slippery, but it isn't. When you skate, the ice beneath your skates' sharp runners melts temporarily, creating a thin lubricating film of water. When it is very cold, the ice does not melt readily; this makes the ice feel "sticky" and harder to skate (or slip) on.

The Asteroid

Zero gravity would prevail at all points inside the hollowed-out asteroid, so a floating object would maintain its position.

Shooting Stars

As the Earth passes through a swarm of meteors its "face"—that is, in the direction the Earth is travelling around the sun—encounters more meteors than its "back" does. For a similar reason, when you run through the rain more water falls on your face than on the back of your head. In the early evening, you are on the trailing side of the Earth and the only meteors you will see are those that "overtake" the Earth from behind. Between midnight and sunrise, you are on the Earth's face and are looking in the direction from which most collisions come. The phenomenon is the same in Australia and everywhere else on the planet.

The Tank

If the tank is suddenly moved to the right, inertia will cause the steel sphere to move to the left relative to the tank. The steel sphere tends to persist in its state of rest. The water tends to do the same, but the sphere, being heavier than water, dominates. Because the cork sphere is lighter than water, it moves to the right relative to the tank. If the tank were moved back and forth, the two spheres would also move back and forth: the steel sphere in the opposite direction, the cork sphere in the same direction.

The Rainbow

At the circle's center is the shadow of your head. The droplets in a rainbow are on the surface of an imaginary cone that points straight to the sun, behind you, and has its vertex at your eyes. (In the case of nearby rainbows, as in a lawn sprinkler, you may be able to see twin, overlapping bows, one from each eye.) A rainbow's arch is always 42 degrees away from the line of sight between you and your shadow. The full circle fills a visual angle of 84 degrees, nearly a right angle.

Field of Play

According to Euclid, a plane can touch tangentially a sphere only at one point, X, as illustrated. Only one person standing at X can be considered upright with respect to the center of the Earth. This is somewhat theoretical as the football pitch is negligibly small compared to the surface of the Earth. Nevertheless, it is perfectly true.

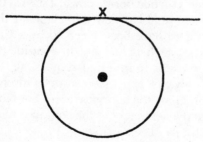

Tumbling Polyhedrons

No. If you could make a polyhedron that was unstable on every face, you would have created a perpetual-motion machine.

Accident in Space

Firing the bullets will propel you by recoil towards the spaceship, if fired in the opposite direction. With each bullet your speed will accelerate.

Suppose the bullets leave the barrel at 1,000 metres per second. Then, depending on the ratio of the mass between the bullet and you, the acceleration could be as much or more than freefall in the Earth's gravitational field (about 32 feet per second^2) and you would not survive the impact with the craft. You would have to decelerate by firing the last bullets in the direction of the spaceship in order to reduce the velocity with which you hit the spacecraft.

Hourglass Puzzle

When the sand is in the top compartment of the hourglass, the high center of gravity tips the hourglass to one side. It is kept at the bottom of the cylinder by the resulting friction against the side of the cylinder. After sufficient sand has flowed into the lower compartment to make the hourglass float upright, the loss of friction enables it to rise to the top of the cylinder.

It is interesting to note that, if the hourglass is a little heavier than the water it displaces, the apparatus works in reverse. That is, the hourglass normally rests at the bottom of the cylinder, and when the cylinder is turned upside down, it stays at the top, sinking only after sufficient sand has flowed to eliminate the friction.

The toy is said to have been invented by a Czechoslovakian glassblower, who made them in a shop just outside Paris. Strangely enough, physicists seem to find it more puzzling than other people. They often advance explanations which involve the force of the falling sand keeping the hourglass at the bottom. However, it is easy to show that the net weight of the hourglass remains the same whether or not the sand is pouring.

Two Bolts

Assume that bolt A has not a thread but parallel rings and you swing bolt B around it. The heads will approach or move away, depending on whether you swing counter- or clockwise. As, however, bolt A has the same thread as B, they cancel each other out.

Another explanation: if instead of swinging bolt B around bolt A, you leave B in situ but turn it clockwise, then the heads will approach. Indeed, this is the function of a screw, as bolt A can be considered a nut. If you turn B counterclockwise, you "unscrew" and the heads will move away. However, if you do neither, but just swing B around, then the heads will remain where they are.

Blowing Out the Candle

The trick is to lower or raise the funnel until the flame of the candle is in line with the upper or lower dotted line. Any attempt to blow out the candle with the flame opposite the center of the opening of the funnel is hopeless, because movement of air is in the boundry layer.

Climbing the Mountain

Bob was most correct when he stated that the cold was due to the rarefied atmosphere. While it is true that the sun's rays are stronger at greater heights and that winds have a cooling effect, and the greater the height the stronger the winds, the main reason for the lower temperature at greater heights is the rarefied atmosphere. With fewer molecules per cubic inch, fewer molecules bombard us each second and less heat is transmitted to us by the surrounding body of air.

Litmus Test

A red rose petal. Press it hard against some paper, and the resultant pink stain will have the same properties as litmus. In fact a rose blossom can be made a deep blue with ammonia.

Sunken Sub

Water pressure pushes perpendicular to a submarine's hull at every point, at the bottom as well as at the top and sides. When a sub settles on a clay or sandy bottom, the water layer may be squeezed out from beneath the hull, robbing the sub of much of its upward buoyant force. In effect, the downward forces can then "glue" the sub to the bottom.

Feathers and Gold II

An ounce of gold weighs more. The Troy system has 12 ounces to a Troy pound, whereas in the Avoirdupois system a pound consists of 16 ounces. So a Troy ounce is greater than an Avoirdupois ounce.

The two Feathers and Gold puzzles demonstrate the vital importance of defining precisely the units in which a measurement is made. Units of measurement are objective, ultimately reflecting some measure in the world as observed by man. Thus, the measure of length is a foot, the measure of the size of a horse is the hand. Both the Troy system and the Avoirdupois system are based on the weight of a grain of wheat as developed by man, which (to use another system, metrics) is 0.0648 grams. A Troy pound is 5,760 grains, and an Avoirdupois pound is 7,000 grains. In the words of the first Greek sophist, Protagoras (fifth century B.C.): "Man is the measure of all things." Whereas Protagoras meant this subjectively, to suggest that judgments are relative, it is better interpreted objectively, to express the nature of units of measurement.

Suspicion

Shortsightedness (myopia) is caused by defective refraction in the eye's lens. In myopia, near objects can be seen clearly while objects in the distance appear blurred because they are focused in front of the retina. Concave lenses change refraction and correct the defect.

Strangely enough, pinholes have a similar effect. Therefore, all you have to do is find a piece of cardboard and make a pinhole. Looking through it, you will be able to read the number plate.

Railways

The rails expand as temperature rises. If the temperature of X feet of rails is raised by T degrees, then the length increases by an amount of 0.000006 XT. On a hot summer's day very large forces would be set up causing the rails to buckle. This expansion used to be accommodated by a gap between every rail section joined as shown. Free movement at the rail joints is allowed for by slotting the bolt holes.

Fishplate

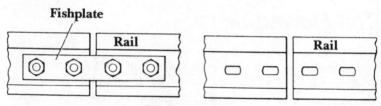

Expansion Joint

Since the 1950s, conventional wisdom has undergone some rethinking. It is now accepted that rails can be welded together into lengths up to half a mile and even longer, without a break. By anchoring the rails more closely to the ties and laying rails at a time when temperature is close to, or slightly in excess of, the mean temperature of the site, it has proven possible to reduce the effects of heat expansion significantly. The residual heat expansion problem is taken care of by planing the end of the long sections of rails and overlapping them.

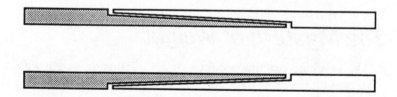

Getting Tanned

Contrary to popular belief, ultraviolet light reflects very little off water. It is the high reflection off sand that contributes most to your increased exposure at the beach.

The Movement

The movement of standing on one's toes, or raising the knees when sitting (keeping the toes on the floor). In the latter case you can lift a very heavy person sitting on your knees with very little sense of strain compared to what you would feel when lifting him with your arms. The calf muscles seem to be the most powerful and efficient in the human body.

The Flawed Sense

The sense we are referring to is sight.

Motion pictures consist of a succession of stills, usually 24 frames per second. The eye perceives these stills as smoothly flowing movements because of a phenomenon called "persistence of vision." The impression of an image on the retina lingers on, as it is retained by the brain for a fraction of a second after the image has disappeared.

Soap Bubbles

The small bubble has more internal pressure than the larger one, so it will shrink as the larger bubble expands. The same principle explains why a balloon is initially hard to inflate, but becomes easier as it expands.

The Mystery of Weight

1. The weighing machine would indicate a reducing weight with increasing speed of rotation.

2. If the Earth's rotational speed continued to increase, then at a certain critical speed, the centripetal force necessary to keep us moving in a circle would be just equal to the total gravitational force. No resultant force would be left over to provide weight and so the weighing machine would read zero. In other words, we have become weightless although the full gravitational force still continues to act.

The Falling Elevator II

If an object which initially has no velocity relative to the elevator is released in the elevator, then this object moves downward with uniform acceleration relative to the elevator shaft, and hence has exactly the same motion as the falling elevator. It thus remains floating in the elevator. Therefore, as a consequence of the breaking of the cable, a situation has arisen in the elevator just as if gravity had disappeared, and as if the bodies no longer had any weight. However, their mass has not been diminished.

The drop of mercury, which was flattened by gravity, resumes its spherical shape since its surface tension still exists. The center of gravity of the drop of mercury moves upwards relative to the elevator and retains the upward velocity thus acquired since there are no forces that reduce this velocity. Hence the drop of mercury moves upwards in its entirety, always relative to the elevator. When it reaches the ceiling of the elevator, it is impelled back down again, and keeps oscillating in this way between floor and ceiling.

The candle goes out for want of oxygen. The carbon dioxide and water vapor formed by the combustion are strongly heated, and, as a result lighter than air—although carbon dioxide is heavier than air of the same temperature. Therefore, the combustion products dissipate in normal circumstances, giving fresh air an opportunity of access. But, in the falling elevator, differences in specific gravity have no effect because gravity has disappeared. Therefore the gases in question remain hanging around the wick of the candle for a long time and spread only slowly by diffusion through the elevator.

Saving Syracuse

Archimedes' feat is entirely practical. It was reconstructed in 1973 by a Greek engineer who had 70 mirrors (each about 5 feet × 3 feet) held by soldiers, who focused the sun's rays on a rowboat that was anchored about 160 feet offshore. A few seconds after the mirrors had been properly aimed, the boat started burning, and was eventually engulfed in flames. In order to work, the mirrors would have to be slightly concave, with the focus on the rowboat.

Bootstrap Elevator

Although it looks as if the man were trying to lift himself up by his own bootstraps, he really isn't. True, for every pound of force that he pulls up on the rope, he also pushes down an equal force on the block, but if he is strong enough to lift his own weight plus the weight of the block, he will rise from the ground. (Tests have shown that a 190-pound man can lift both himself and a 110-pound block in this way.)

The Mirror Phenomenon

Left and right are directional concepts while top and bottom, or up and down, are positional concepts. The same incidentally applies to east-west and north-south. Walk northward along the Greenwich Meridian and Berlin will be to the east and on your right (ignoring the fact that you could also travel to Berlin the long way round) until you reach the North Pole. Crossing it, you turn around to still look north and Berlin will be to the west and on your left. Yet north and south will remain in the direction of the poles. Equally, up and down assume the center of the Earth as reference point. An ordinary mirror will reverse direction, but not position.

Center of Gravity

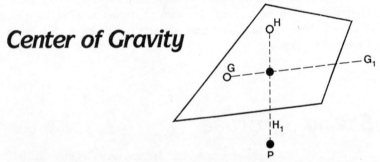

You drill a hole, H, in your template and suspend the box so that it can swing freely. The center of gravity is clearly somewhere on the vertical line H–H1. This line can be marked by a thread, and a weight, P, suspended from H. This procedure is then repeated through hole G. The point of intersection of H–H1 and G–G1 is the center of gravity.

And Yet It Moves

Experiment 1: Weigh yourself on the Equator and then on one of the poles. You will weigh slightly less on the Equator because the centrifugal force acts against gravity. This is not the case at the poles.

Experiment 2: Use a train travelling in the northern hemisphere towards the North Pole, starting from the Equator. If, as we want to prove, the Earth rotates, then the easterly velocity will be maximum at the Equator, decreasing as we travel north, becoming zero at the North Pole.

But the train itself would try to retain the maximum easterly velocity by virtue of the phenomenon of inertia. The rails prevent the easterly motion, and because of this the right-hand rail is subject to increased pressure, which could be measured.

Experiment 3: Drop an object into a deep mine shaft. For a depth of 100 metres, the easterly deviation is 2.2 cm at the Equator.

I hope you are now satisfied. The effect of the Earth's rotation is also evident in the phenomena of trade winds and ocean currents.

Gravity

The moon's most pronounced orbit is around the sun, just as the Earth's is. So, in a sense, the sun has captured the moon. The Earth causes only a small deflection in the solar orbit which the moon would have if the Earth weren't here.

The Fast Car II

You would have to travel the second half of your trip at infinite speed.

Evolution

A spherical or oval egg would roll in a straight line and would be more likely to fall out of the nest, if built, for instance, on a cliff-edge. However, as an asymmetrical shape, the egg would roll in a circle.

Iron Doughnut

As the doughnut expands, it keeps its same proportions; so the hole also gets bigger. This same principle is at work when an optician removes a lens from a pair of glasses by heating the frame. The next time you can't open the metal lid on a stubborn jar, heat the lid under hot water. The lid, inner circumference and all, will expand, making it easier to loosen.

The Spectrum

Pigments in common use are impure. Because of these impurities, yellow paint presents the eye with a mixture of red, yellow, and green, and blue paint offers a mixture of blue and green. In each case, however, the basic color dominates so that the eye perceives only yellow and only blue respectively. However, when the two paints are blended, yellow paint absorbs blue, and the blue paint absorbs red and yellow light, leaving green as the only remaining color common to both paints.

The colors of projected light, however, are pure, and if they are complementary colors, such as blue and yellow, or green and red, white light is the result.

In other words, when paints are mixed, the resulting color is produced by absorption. When lights are mixed, the results are produced by combination.

Quito

He would weigh zero. His center of gravity would be in orbit, considering the Earth's speed of rotation and the fact that Quito is on the equator.

Vacation on the Moon

1. Rocket ships are in free-fall as soon as they leave the Earth's gravity field. From the time the motors are turned off to the time they are used again for altering course or braking, there is zero gravity inside a rocket ship.

2. Cigars won't stay lit in zero gravity unless you constantly wave

them about, for the reasons explained in the solution to "The Falling Elevator II."

3. Birds can't fly on the moon because there is no air against which their wings can push or support them when gliding.

4. No air, no breezes, no rippling flags on the moon.

5. Although in daytime the lunar sky is indeed dark, there is so much reflected light from the moon's surface that stars are not visible to unaided eyes. They *can* be seen through binoculars.

6. Even at night, stars on the moon never twinkle. Twinkling on Earth is caused by movements of the atmosphere.

7. For stars to be visible inside the arms of a crescent Earth they would have to be between Earth and the moon.

8. The moon does rotate once during each revolution around the Earth, but since it always keeps its same face towards the Earth, the Earth does not rise and set. From any given location on the Earth-side of the moon, the Earth remains fixed in the sky.

9. Without air a boomerang can no more operate on the moon than a bird can keep itself aloft.

10. Twitchell couldn't have heard the boomerang strike the boulder because sound requires an atmosphere to transmit its waves to a human ear.

11. Before the first moon landing it was widely thought that objects would be invisible in moon shadows. Actually, so much light is reflected from the irregular lunar surface that this is not the case.

12. Although the sun does rise and set on the moon, it takes it about 28 days to return to a former position. It could not have set as rapidly as the narrative indicates.

13. The terminator moves at about 10 miles per hour. This is slow enough for a person to keep pace with its movement.

14. Meteors leave glowing trails only when they are burned up by friction of the Earth's atmosphere. On the atmosphere-less moon, meteors would not produce such trails.

15. As in mistake 12, the sun could not have risen until some two weeks after it set.

Balloon Behavior

As the car accelerates forward, the balloon on its string tilts forward, too. Inertial forces push backwards in the car, pressing the people against their seats (an effect with which we are all familiar), but also compressing the air at the back of the closed car. This increased air pressure at the rear pushes the balloon forward. For similar reasons, as the car rounds a curve, the balloon tilts into the curve.

The Half-Hidden Balance

From one pound to infinity. It is 1 pound, not 2, because when the length approaches infinity the second weight converges to zero. When the hidden rod's length is zero the weight must be infinite to balance the 1 foot–pound force on the left.

Balance the Beaker

The answer is 2 cubic centimetres of water must be added to the left-hand beaker: one to balance the cubic centimetre of water added to the right-hand beaker and one to balance the buoyant force exerted on the iron bar due to its displacing one cubic centimetre of water when the water level has been raised one centimetre.

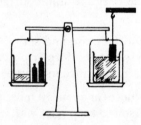

The Wheelbarrow

Pulling a wheelbarrow is easier. Pushing presses the wheel harder against the ground, adding to the workload.

The Bottle and the Coin

Isaac Newton's first law of motion explains this experiment:

Every body continues in its state of rest or of uniform motion in a straight line, unless compelled by some external force to act otherwise.

This tendency to continue in its state of rest or uniform motion is called "inertia."

Several other tricks are based on the same principle. For example, if a pile of coins is placed on a table, the bottom one can be removed without disturbing the remainder simply by flicking it sharply with a piece of thin wood or metal. In both cases, the incidence of friction is insufficient to overcome inertia.

Big into Small

If the unpeeled egg is soaked overnight in vinegar, or acetic acid, its shell becomes plastic. The procedure described will put the egg in the bottle, and a cold-water rinse will restore the shell to its original hardness.

Skipping Stones

The first skip is short and to the right. When the stone's trailing edge hits, it pushes sand to the left; the stone tilts forward and hops to the right. Then the leading edge strikes and pushes sand to the right; the stone tilts back and takes a long hop to the left, and the cycle repeats. The short hops appear to be missing when stones skip over water. After the trailing edge strikes, the stone planes along, building a crest of water in front of it, then lifts out and makes a long hop. It strikes with its trailing edge again and repeats the process.

The Mix

A is already boiling. Further heating will merely reduce its quantity without raising its temperature, so there will be less to mix with B.

A Maritime Problem

It all depends on how the oval wheels are affixed to the vehicle's axles. If the wheels on the opposite ends of the same axle are positioned at right angles to each other, a roll will be produced. By synchronizing the front and rear wheels, so that on each side of the vehicle the two wheels have their long axes at 45-degree angles, the carriage will both pitch and roll. If, on each side of the vehicle, the two wheels also have their long axes at right angles, the carriage will merely move up and down alternately on two diagonally opposite wheels.

These possibilities leave only the problem of finding a driver prepared to put up with any of them.

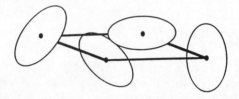

Free-Fall

The cork will still be suspended in the middle of the bucket because, in free-fall, buoyancy ceases to exist.

Bank Note

No actual picture is made of it except on the television screen, where it is evanescent. The recording on tape is a linear message or code, not a recognizable picture.

Sand on the Beach

Before you step on it, the sand is packed as tightly as it can be under natural conditions. Your weight disturbs the sand, making the grains less efficiently packed. The sand is forced to occupy more volume and rises above the water level, becoming dry and white. The water rises more slowly, by capillary action, so it takes a few seconds or more before the sand gets wet and dark again.

Athletics

About 18.5 feet. It isn't the distance from the ground to the bar that is important, but the distance from the jumper's center of gravity (CG) to the bar. A six-foot male's CG is about 3.5 feet above the ground; so to jump his height he must raise his CG 2.5 feet. On the moon he would be able to raise his CG six times as high, or 15 feet. By raising his legs up to clear the bar the same way he does on Earth, he could add that 3.5 feet back to the jump (just as on Earth), bringing him to the 18 feet 6 inches mark, which is just over half the height that is commonly quoted.

The Brick and the Dinghy

Throwing a brick into the dinghy. In the water the brick displaces its volume of water; in the dinghy it displaces its weight of water. Since a brick is heavier than water, it weighs more than its corresponding volume of water does.

The Spinning Top

This inversion isn't easy to explain, but various theories have appeared. Jearl Walker, in *Scientific American*, says that the simplest explanation is that the flip-over "arises from friction between the top and the surface on which it spins." That is, whenever the spin axis tilts away from the vertical, the top slides on part of its spherical surface. The friction creates a torque that precesses the top to an inversion. Even with that explanation, this simple toy nevertheless seems to violate common sense.

The Tippe Top raises a second paradox that isn't often discussed. If the top is spinning clockwise at first, as seen from the stem side (above), it will spin counterclockwise, as seen from the stem side (below), after the inversion.

If you could continuously view the flat part of the stem, the top's rotation would be seen to slow down and stop at some point during the inversion and then to start up again in the other direction! It may seem amazing, but it's true.

The Camargue Horses

The herds were plagued by vicious horseflies, causing severe debilitation of the animals. For some inexplicable reason, the flies showed a strong preference for the darker horses and did not attack the white ones, resulting in the survival of the famous "white horses of the Camargue."

Hole Through the Earth

1. The ball's velocity would steadily increase from zero at point A to a maximum at the center of the Earth. It would steadily decrease thereafter to zero at point B, taking 42 minutes for the complete trip. This fascinating speculation goes back to Plutarch. Even Francis Bacon and Voltaire argued about it. Galileo gave the authoritative answer which is generally accepted:

The ball would fall faster and faster, though with decreasing rate of acceleration, until it reached maximum velocity, about 5 miles per second, at the Earth's center. It would then decelerate until its speed reached zero at the far end of the hole. If air resistance is ignored, it would oscillate back and forth, like a pendulum, *ad infinitum*.

2. At the center of the Earth, there is no overall gravitational force on the ball. This is because the Earth's pull on the ball will be equal in all directions. Therefore, the weight will be zero.

3. The weight will change but the mass will not.

4. The ball would be in free-fall throughout the entire trip, so it would always be in a state of zero gravity.

5. More time. The trip would take about 53 minutes. Although the distance is much shorter than on Earth, the moon's gravity is only about one-sixth of that of the Earth.

Rotation Counter

In riding, the front (steering) wheel follows a more wobbly, therefore longer, course than the rear one.

Drops and Bubbles

The bubbles would attract each other. If water is removed from one spot in all space (bubble A), the gravitational balance surrounding it is upset, and the net effect on a nearby molecule of water is that it is drawn toward greater mass; that is outward, away from the bubble. If there are two bubbles, the water between them acts as if it is repelled from both, and the bubbles would move toward each other.

The Smooth Table

If I throw an object in a horizontal direction, the common center of gravity of myself and this object will still remain on the same vertical line. Thus, if I throw the object, for instance my jacket, to the right, my center of gravity will move to the left, although more slowly than the object, assuming it has smaller mass. Slowly I will slide off the table, even if I only exhale air by blowing. In this case I would effectively act like a rocket.

Earth and Moon

The Earth's rotation is gradually slowing down. Don't look forward to a 25-hour day, or to getting a few extra minutes of sleep in the morning, however. The effect is just enough to add about one second to the Earth's day every 100,000 years.

Rays of the Sun

The sun's rays are parallel, of course. The "fanning out" is an optical illusion, the same illusion that makes railroad tracks appear to meet at a point on the horizon. These rays are a more powerful illusion because of the absence of reference points. It's hard to believe, even when you know it for a fact, that two rays high overhead and "so far apart" are just as far apart in the distance where they first "emerge" from behind the cloud.

Cool Off

You might cool off temporarily, but the net effect of leaving the refrigerator door open will be to make the room hotter. The refrigerator's cooling system will activate in an attempt to cool the refrigerator air again, and more heat will be released by the motor than can be absorbed by the released cool air, so the apartment will become even hotter.

Today's New Yorkers are only too aware of this effect, complaining that, in summer, subway platforms seem to be even hotter than they used to be. Since air-conditioned subway cars were introduced, they are! The air-conditioning systems' motors give off more heat when a train is in a station than the cooling effect of the cold air escaping through open doors.

The Helicopter

Every revolving body develops a force acting perpendicular to the axis of revolution. This moment of force is referred to as torque, and would cause the helicopter to rotate out of control. The tail rotor is designed to counteract the torque.

In large helicopters with two rotors, the torque is avoided by having the rotors rotate in opposite directions.

The Falling Elevator I

Before the free-fall, the spring is stretched by the cork's buoyancy, equal to the weight of the water displaced by the cork. In free-fall

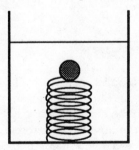

there is no weight, only mass, and the spring contracts to its natural state, pulling the cork (which has lost its buoyancy) with it.

Table and Box

The first box contains uncooked rice piled at an angle (as illustrated, left). The second box contains an iron weight and a pile of ice. It has to be finely balanced, so that the weight of the ice keeps the box on the table. In due course the ice will melt, toppling the box.

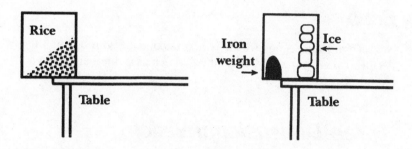

Thermal Expansion

The initial fall to level B–B is due to the expansion of the flask, which is heated and expands before the heat has had time to act upon the water. Once the glass of the flask has reached the limit of its expansion, the water in the tube begins to rise.

The Space Station

1. If the astronaut walked round the station in the direction of its rotation his weight would increase. If he turned round and walked in the opposite direction his weight would decrease. Why should this be? We know that the centripetal force increases proportionally with the speed of rotation. By walking in that direction the astronaut adds his speed to that of the rotation.

2. Try to spin one of your coins on the floor of your room: the coin will refuse to spin. By conservation of angular momentum, a spinning object tries to maintain its position in space. Since the spinning station is continually changing your position in space, a coin that is spun will keep changing its orientation, to correct its angular momentum, and will topple and fall.

Gravity Train

The gravity train would indeed work. Strangely enough, such a train would make all of its trips in about 42 minutes, the same time it would take an object to fall through the center of the Earth. The time is constant regardless of the tunnel's length.

Echo

No recognizable echo, which is a reflection of sound, but the decibels (intensity) would be doubled, much as is candlelight when placed next to a mirror.

Three-Dimensional Vision

The phenomenon of 3-D vision rests on the fact that the two eyes perceive slightly different versions of the same object, as shown in the illustration.

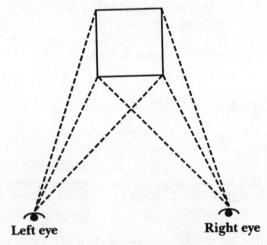

Left eye   **Right eye**

The human brain is genetically programmed to integrate the two views of the same objects into three-dimensional vision. If you close one eye, the three-dimensional world you see around you becomes two-dimensional. You might not see much difference, but that is because your visual memory knows that what you are looking at is, in fact, three-dimensional and persists in "seeing it" that way. A one-eyed person, however, is unable to see in three dimensions so cannot estimate distances, an ability that relies on the processing of different images received by the two eyes.

Any attempt to make a two-dimensional image appear three-dimensional relies on deceiving the brain into reacting as if each eye were receiving a different image of the same object.

Years ago, the stereoscope, an optical instrument with a binocular eyepiece, presented two slightly different pictures or photographs for viewing, one with each eye. The brain was thus deceived and offered the illusion of a three-dimensional picture. The short-lived 3-D movie presented overlapping images in red and green that were filtered out by the red- and green-lensed spectacles, again fooling the brain. These 3-D illusions, and some others, are

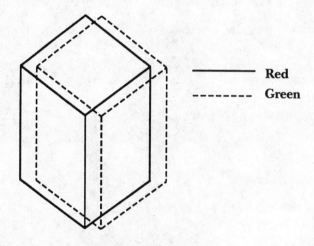

————— Red

- - - - - - - Green

based on the fact that human eyes usually focus on whatever object they are looking at. However, the eyes can defocus, or diverge, and this is the principle behind stereograms.

Let's do an experiment. Hold a pencil vertically right in front of your nose, relax, and then slowly move the pencil about fifteen inches away from your nose. You will most likely see two pencils, unless you try deliberately to focus. Now follow through with two further experiments which will lead to a full understanding of the stereogram phenomenon.

First, close one eye and then the other, alternatively. You will find that first one and then the other pencil will disappear. Second, turn the pencil to a horizontal position. The two pencils will merge into one, except for the two ends.

In a stereogram the patterns are duplicated, but with minute differences which are not observable because of the background. If you let your eyes wander, i.e., diverge or defocus, each eye will pick up a different version of the duplicated pattern. That is all the brain needs to produce 3-D vision. You can prove this theory with any stereogram pattern, such as the one below. Once you see the image in three dimensions, close one eye and the 3-D illusion will disappear. The same will happen if you turn the image sideways right or left, but it will reappear if you turn it upside down (180 degrees).

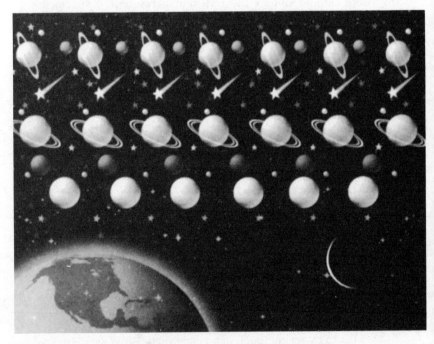

Black-and-white reproduction of artwork from *Magic Eye II*, © 1994 by N. E. Thing Enterprises. Reprinted by permission.

Shut That Door

Nothing would happen because there is no oxygen in the room.

Index

About the Author

Erwin Brecher was born in Budapest and studied mathematics, physics, engineering, and psychology in Vienna, Brno (Czechoslovakia), and London.

He joined the Czech army in 1938 and, after the Nazi occupation of Sudetenland, he escaped to England. Engaged in aircraft design during the war, he later entered the banking profession, from which he retired in 1984. Currently, he devotes his time to playing bridge and chess, and to writing books on scientific subjects or of puzzles, such as his recent *Lateral Logic Puzzles. Surprising Science Puzzles* encompasses several of his interests.

A member of Mensa, Erwin Brecher makes his home in London, England.